UNLOCKING ECLIPSES

WHAT TO LOOK FOR, HOW TO VIEW THEM, AND WHY THEY MATTER

Dallas, Texas
ICR.org

UNLOCKING ECLIPSES
WHAT TO LOOK FOR, HOW TO VIEW THEM, AND WHY THEY MATTER

First printing: August 2023

Content developed by Bethany Trimble, Renée Dusseau, and Rachel Brown
Contributions by the ICR Communications team: Jayme Durant, Susan Windsor, Beth Mull, Michael Stamp, Lori Fausak, and Frank Sherwin

All Scripture quotations are from the New King James Version.

ISBN: 978-1-957850-28-3

Please visit our website for other books and resources: ICR.org

Printed in the United States of America.

Published by Wayfinders Press, an imprint of ICR Publishing Group

TABLE OF CONTENTS

Time-lapse of total solar eclipse

INTRODUCTION

Eclipses have captivated people throughout history with their breathtaking displays of shadow and light. They are stunning examples of Christ's magnificent handiwork, and their splendor proclaims the majesty of the infinitely glorious Creator.

The book of Genesis tells us that in the beginning God created the heavens and the earth (Genesis 1:1). He placed all the stars exactly where they need to be and "calls them all by name" (Psalm 147:4). In His creativity, Jesus gave each celestial object unique features and placed humans on the planet specially situated for us to marvel at the incredible events in our intricately designed solar system—such as eclipses.

Today, scientists know more about these occurrences than ever before. Modern technology has allowed researchers to closely observe the processes that cause eclipses, both on Earth and beyond. Sadly, many people believe these events are the result of mere coincidence in a universe randomly spun into existence by a Big Bang billions of years ago. In this view, humans are simply lucky to be alive at the right time and on the right planet to witness them.

What does the evidence actually tell us? In truth, eclipses point to a purposefully engineered universe. Every astronomical detail—from the orbit of planets to their size, location, and gravitational pull—comes together to form a stunning revelation of beauty and order. Perhaps even more wonderful is that we have a front row seat to see it.

In this booklet, we'll discuss the science behind eclipses. We'll also provide a guide to maximize your eclipse day experience, including safety tips and FAQs. Most importantly, we'll discover how the precisely orchestrated mechanisms of eclipses, objects in space, and even space itself indicate the perfect wisdom of their Maker.

Psalm 19:1 says that "the heavens declare the glory of God; and the firmament shows His handiwork." As you learn about eclipses, meditate on how they reflect the greatness of our King. His marvelous work is clearly seen through the exact properties and alignments of the moon, sun, and Earth to create these wondrous sights—all for the glory of our Lord and Savior, Jesus Christ.

Did You Know?

The term eclipse comes from the Greek word *ekleipsis*, which means "an abandonment." It's related to a verb that translates "to forsake, fail to appear."

PART ONE: THE BASICS

In the past, eclipses inspired many colorful myths. In ancient China, records describe the sun being eaten. The legends of one Native American tribe speak of a bear fighting the sun, while others interpret eclipses as the result of an angry sun.

But what's really behind these unusual events? The explanation might be simpler than you think. An eclipse happens when two objects in space are aligned so that one object casts a shadow upon the other. Explained a different way, one space object totally or partially obscures another. On Earth, we're able to see two different types of eclipses—lunar and solar.

In a lunar eclipse, the earth moves between the sun and the moon, casting its shadow on the moon's surface. From our perspective, the moon often takes on a reddish hue. This coloring is a result of the earth's atmosphere. When sunlight hits small particles in the atmosphere, blue-green light is scattered so that only red light reaches the moon and is reflected back to Earth.

A solar eclipse occurs when the moon comes between the sun and Earth, so Earth is in its shadow. From our viewpoint, the moon appears to drift over the sun, blocking its light and causing darkness to fall over part of the earth. The regions that are most affected by the eclipse experience a temporary nighttime. Other areas are dimmed but do not get as dark. Instead, they see only glimpses of the moon cutting off the sun.

Lunar Eclipses

There are three kinds of lunar eclipses. In a **total lunar eclipse**, the moon is located in the center of Earth's shadow, also called the umbra. The moon will darken slightly, turning copper as sunlight filters through the earth's atmosphere. This phenomenon is known as a blood moon. If the air surrounding Earth is especially clouded, the moon's color looks even more vibrant.

When the moon travels across only a portion of Earth's umbra, there is a **partial lunar eclipse**. While part of the earth's shadow falls on the moon, the shadow never fully covers its surface. Therefore, sections of the moon darken before they're brought back into the sunlight.

It can be difficult to see a **penumbral eclipse** because the moon may not appear to be any different. This occurs when the moon moves through the outer region of Earth's shadow, or the penumbra. Because only a small amount of sunlight is blocked, the moon looks slightly darker than usual but otherwise shows little sign of change.

Blood moon

Solar Eclipses

There are four kinds of solar eclipses. During a **total solar eclipse**, the moon travels between the sun and Earth and seems to entirely cover the sun's face. It casts a shadow on portions of the earth, darkening regions that are close to its center. On clear days, the sun's atmosphere, or corona, can be seen encircling the eclipsed face of the sun.

The moon looks like a dark circle just in front of the sun during an **annular eclipse**. This effect takes place when the moon is at its farthest location from Earth, or close to that point. It may appear as though the moon has a bright ring around its circumference.

In a **partial solar eclipse**, the sun looks like a crescent because the moon doesn't completely block its light. Since the moon, sun, and Earth aren't precisely aligned, only a portion of the sun's surface is concealed. These eclipses vary in size, and they're what people see in places outside the moon's umbra during a total or annular eclipse.

When the eclipse changes between total and annular, it's called a **hybrid eclipse**. As the moon passes over the sun, it switches types because of Earth's curvature. These eclipses are rare and are often witnessed only a handful of times in a century.

Total solar eclipses have five stages.

First contact—The moon begins to cover the sun. The sky darkens as lines called shadow bands ripple across the ground. These waves are likely caused by the bending of light rays in the earth's atmosphere.

Second contact—The sun is shielded by the moon. Packets of sunlight known as Baily's beads may appear around the moon's circumference. Just before the sun disappears, one final burst of light can be seen in what's called the diamond ring effect.

Annular solar eclipse

Baily's beads

Totality—The moon covers the sun's entire surface. As the sky darkens, other celestial bodies like planets become visible. The sun's corona, typically hidden in daylight, appears as a halo around the moon. Pink flames or prominences, jets of gas that leap from the sun's surface, might also be seen during a total eclipse.

Third contact—The sun reappears as the moon begins to move away.

Fourth contact—The moon no longer masks the sun, marking the conclusion of the eclipse.

Did You Know?

During a solar eclipse, the course of the moon's shadow as it moves across Earth is known as the path of totality. It's usually only about 10,000 miles long and 100 miles wide, which is less than 1% of Earth's surface.

Eclipse Cheat Sheet

	Lunar	Solar
Cause	The earth moves between the sun and moon	The moon moves between the sun and Earth
Time	Night, during a full moon	Day, during a new moon
Appearance	Reddish-orange moon	Darkened sky as the moon covers the sun
Duration	Several hours	A few minutes
Frequency	Up to three per year	Two to five per year on average
Supplies for viewing	Telescope, binoculars, camera with tripod	Solar filters (found in specialized glasses and handheld solar viewers) or projector
What to look for	Blood moon	Shadow bands, Baily's beads, diamond ring effect, corona, prominences, celestial bodies (planets, stars, etc.)

PART TWO: OBSERVATION DAY

Eclipse days are exciting, and it's important to plan for the event. Where are the best viewing spots? How will you get there? What will you need at the eclipse site, especially knowing that you may be outside for several hours? The following tips offer advice to help you make the most of an amazing eclipse experience.

Tip 1: Safety First

The sun is crucial for survival, but its powerful rays can be harmful. It's never safe or healthy for your eyes to gaze directly at the sun without the correct eye protection, even if it's partially blocked by the moon in a solar eclipse. The light-sensing area of your eye, called a retina, can be permanently burned.

To view a solar eclipse safely, you'll need glasses or handheld viewers with solar filters. You may also choose to observe the eclipse indirectly through a pinhole projector, which is easy to make using supplies you might already have on hand. (See the instructions at the end of this section.) It's also important to use a solar filter for your camera lens, binoculars, or telescope. This protects your device—and retinas—from sun damage.

Don't take chances. Without protective eyewear, the sun can cause severe eye injury or blindness during solar eclipses. For lunar eclipses, though, special filters aren't necessary. Set your alarm, find a spot, and enjoy!

Tip 2: Start Planning Early

Eclipses aren't incredibly rare—several can happen in a year. But total solar eclipses are less common. They're only seen in any given spot on Earth once every few hundred years. Observing a total eclipse is significant because the next one won't occur in that same location for a very long time. As a result, they often attract large crowds, so seeing one may require more preparation than you might expect.

Start by researching the areas where an eclipse will be the most visible. Then, pick a specific viewing spot, ideally a wide-open space where large buildings, trees, or other structures won't block the view. Look for a place without bright lights, which would otherwise ruin the effect of a lunar eclipse when it gets dark. Don't forget to double-check the time frame of

the eclipse and take off work in advance if needed. For a lunar eclipse, plan on setting an alarm (or multiple ones if you're a heavy sleeper). If you plan to travel a significant distance from home, consider booking a hotel or other accommodations, and make sure to secure your room early. The reservations will fill up fast!

Keep a running list of supplies you'll need for observation day. For a solar eclipse, you'll want to pack sunscreen, snacks, water, and solar filters, of course. Because you may be waiting for several hours, it's a good idea to bring a blanket, hammock, or chairs, as well as games to pass the time. Arriving early at the site provides extra time to find the best spot, locate the nearest restrooms, and set up a telescope or camera equipment if desired. Be sure to account for the increased traffic surrounding the event as well.

Before you go, keep an eye on the weather and be prepared to adjust plans as needed. An eclipse won't differentiate between clear and cloudy skies, so you'll have to find another spot—ideally still on the path of totality for a solar eclipse—if haze threatens to disrupt your view. Depending on the time of year, it may also be slightly chilly. If the eclipse falls during the winter months (especially if it's lunar), plan on wearing warm clothes and packing a blanket, thermos, and hand warmers.

Did You Know?

To observe a total solar eclipse in all its splendor, make sure you're fully situated in the path of totality. Even just a few miles outside the path won't result in the same effect.

Eclipses are awe-inspiring on their own, but friends or family can make them even more special. Invite your loved ones, or consider attending an eclipse viewing party. Some communities host festivals that can last several days, and many of them are free. If your hometown falls within the path of totality, you may even consider hosting your own!

Tip 3: Soak It All Up

The day has finally arrived. The spot is secured, the supplies are gathered, and the skies are clear. Now what?

One simple way to remember an eclipse day is through pictures. Capture photos and videos of your surroundings, including people, nature, and anything else worth documenting—but not the sun unless you are using a solar filter.

While photography is often more difficult in the darkness of a lunar eclipse, gazing through a telescope or binoculars may be equally memorable. Unlike a solar eclipse, no special filters are necessary, and the viewing time is longer by several hours.

An additional option is to journal during the eclipse, which allows you to document the experience in a more personal way. If you're interested in the scientific effects of an eclipse, you may also choose to record data with a thermometer or light meter to track changes in temperature and light, respectively.

Most importantly, savor the eclipse as much as you can. Place your devices to the side and take a few moments to simply admire the beauty of Jesus' creation beyond our planet. There's no picture or image that comes close to observing the wonder of an eclipse for yourself.

Eclipse Photography

It's generally recommended to leave eclipse photography to professionals. An eclipse can be difficult to capture, and attempting to take photos can easily distract from the event itself. Each eclipse experience is unique. Once missed, it can't be repeated.

However, there are several tips to increase the efficiency and effectiveness of an eclipse photoshoot if you choose to have one. The following list isn't exhaustive, but it provides a good start.

Never point the camera at the sun without a solar filter except during totality. Failure to do so may damage the light sensor. For lunar eclipses, filters aren't necessary.

Use a tripod to stabilize the camera. Taking pictures while holding the device may result in camera shake. To further reduce vibration, **delay the shutter time** before taking the photo and consider using either a **cable or electronic shutter release**.

Don't use flash. It won't add to the quality of your photograph, and it can distract the people around you.

Set the camera to manual focus. More specifically, switch the camera to infinity focus, which will provide clearer and higher-quality images of distant objects.

Choose the correct lens. To see eclipse details in the photograph, use a high-zoom lens on your camera—typically with a focal length of 400mm or greater.

Take pictures using the RAW setting if you can. This format provides greater flexibility when editing photographs later.

Don't use a cell phone except to take pictures of your surroundings. The resolution of a phone camera can't compete with the capabilities of DSLR, mirrorless, or other digital cameras. Polaroid and film cameras can also be used to photograph an eclipse.

Try bracketing. Take photos at different shutter speeds during the eclipse to provide several options for editing.

Use a photography mapping app to determine the location of the sun or moon on eclipse day. For serious photographers, **consider purchasing an equatorial mount** that allows your camera to follow the sun or moon as it moves across the sky.

Practice. Taking pictures on eclipse day will go more smoothly if you're familiar with the settings beforehand. Research exposures that other eclipse photographers have used, and be prepared to quickly adjust throughout the eclipse.

Safety FAQs

Can solar eclipses ever be viewed without eclipse glasses?

During a partial or annular solar eclipse, proper eyewear is necessary at all times. However, total eclipses can be observed without eye protection during the very brief time that the moon completely covers the sun. Glasses must otherwise be worn in the phases before and after totality.

What are eclipse glasses?

Eclipse glasses are specialized eyewear made with solar filters that conform to a worldwide safety standard known as ISO 12312-2. They're sold in retail stores and online, but be sure to purchase them early—they can sell out quickly leading up to a total eclipse. Note that it's never safe to use scratched or damaged glasses to view an eclipse.

Why can't sunglasses be used? Aren't they the same thing?

Even sunglasses with 100% UV protection aren't safe to use

during eclipses. They're not built to protect the retina from all the damaging rays. Only eclipse glasses with a special filter are designed for this purpose, and they're many times darker than traditional sunglasses. Warning: Don't try to use multiple pairs of sunglasses together as a filter. They aren't designed to view the sun directly.

How can I tell whether my glasses are truly safe?

Unfortunately, there are many counterfeit glasses available that don't offer sufficient protection. It can be easy to mistake them for the real deal, but thankfully there's a way to check if your glasses are trustworthy. Make sure your glasses are labeled with the ISO icon and reference number 12312-2. It's also strongly recommended that you verify the manufacturer. There are several companies that have been certified by the American Astronomical Society and NASA as reputable, so review this list before purchasing glasses from an unknown supplier.

Should I keep my prescription glasses on during the eclipse?

Yes. Place your eclipse glasses over the prescription pair during a solar eclipse.

Are there any alternatives to eclipse glasses to observe a solar eclipse safely?

Yes! If you can't get eclipse glasses, projectors can work just as well. They allow you to watch a solar eclipse indirectly. See below for how you can make one.

How to Make a Pinhole Projector

A pinhole projector projects an image of the sun onto a nearby surface. It's a great way to view the eclipse safely since no special eyewear is required. To make your own, you'll need two notecards and a thumbtack or sharp pin.

1. Poke a hole in the middle of one notecard with a thumbtack or sharp pin.
2. With your back against the sun, hold the hole-punched card in one hand and the blank card in the other.
3. Lift the first card above your shoulder so that light reaches the pinhole and projects onto the second card.
4. Hold the two cards farther apart to make the reflected sun appear larger.

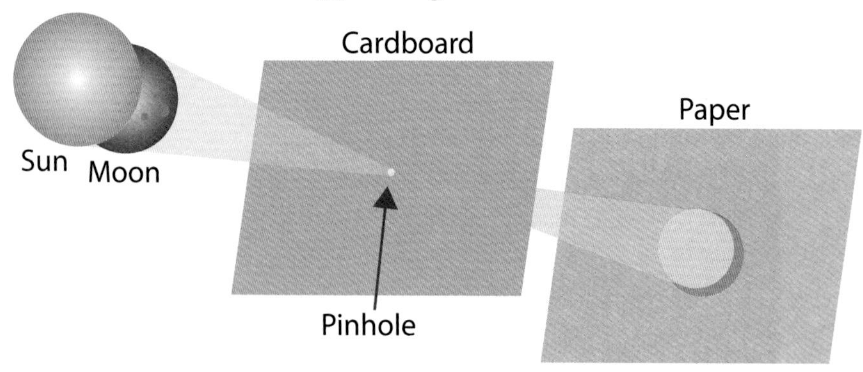

How to Make a Cereal Box Pinhole Projector

A more durable option is a cereal box projector. The supplies you'll need are an empty cereal box, tape, scissors, white paper, a pencil, aluminum foil, and a pin or thumbtack.

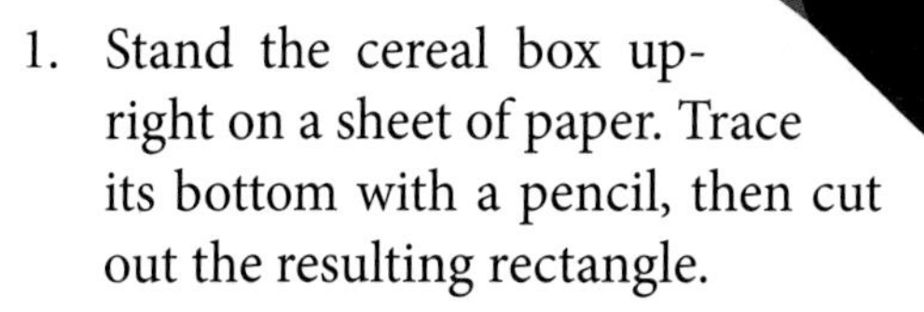

1. Stand the cereal box upright on a sheet of paper. Trace its bottom with a pencil, then cut out the resulting rectangle.
2. Place tape on the paper rectangle, then put the rectangle inside the box and attach it smoothly against the bottom.
3. Seal the box's top so that it lies flat. Then, make two rectangular holes at either end of the sealed top by cutting along its edges. The size of the holes doesn't matter as long as they're about the same dimensions and have cardboard in between them.
4. Tape a piece of aluminum foil over one of the openings in the top, making sure it isn't crinkled and it fully covers the hole. You'll look inside the

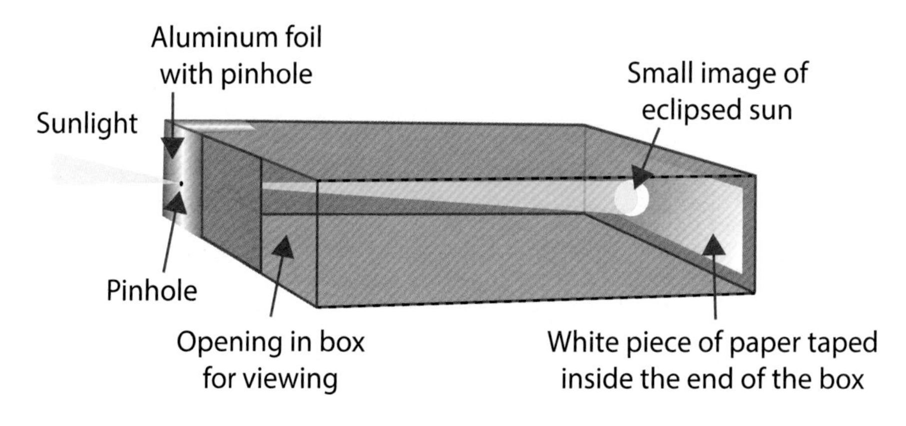

box through the other hole to view the eclipse projection, so leave it uncovered.

5. Poke a small pinhole at the foil's center using a pin or thumbtack.
6. Turn your back to the sun and hold the cereal box so that light can reach the pinhole. While you gaze through the open rectangle at one end, sunlight will filter through the pinhole at the other and project an image of the eclipsed sun onto the white paper inside the box.

Note: Using a taller box will make the reflected sun appear bigger.

How to Use a God-Given Projector

If you don't have a pinhole or box projector on eclipse day, there's a natural projector that also works as a safe option. Simply find a nearby tree, stand underneath, and look at the ground. The light filters through spaces where leaves don't overlap, creating "pinholes" for projections of the eclipse to poke through.

PART THREE: GUIDE TO THE SKY

Eclipses are one of the many wonders available to view across our vast sky. There's so much yet to be discovered about our incredible universe, and we see only a small percentage of it from our limited viewpoint on Earth. Yet, everything we observe—and even what we don't—was designed with purpose. Let's take a look at some of the marvels within our created cosmos.

Earth

Our terrestrial home is the third planet from the sun, and it's perfectly engineered for life. Earth is known as the "Goldilocks planet" because it offers the precise conditions necessary for creatures to live.

Located 93 million miles from the sun, our planet maintains a stable environment that's not too hot or too cold. If it were slightly closer to the sun, the polar ice would melt, and sunlight would bake our planet. If it were a bit farther, large portions of Earth would freeze and become uninhabitable. Our planet has been purposefully placed in the solar system's habitable zone, allowing liquid water to exist and maintaining a temperate climate for life to thrive.

Pillars of Creation, part of the Eagle Nebula

The even distribution of sunlight across our planet is also a crucial factor for life on Earth. In the 17th century, the Christian astronomer Johannes Kepler realized that all planets move around the sun in stretched-out loops called ellipses. Before then, most scientists thought that the planets traveled in circles.

The idea that planets orbit the sun was still a new concept. In Kepler's day, many believed these celestial bodies circled Earth instead. The discovery of the elliptical orbit is now known as Kepler's first law of planetary motion. Earth moves in a near-perfect circle around the sun, providing it with equal sunlight all throughout the year.

> "This most beautiful system of the sun, planets, and comets, could only proceed from the counsel and dominion of an intelligent and powerful Being."
>
> — Sir Isaac Newton

Even the parts of Earth that aren't directly visible to us keep us comfortable in our planetary orbit. Thanks to the atmosphere, we have oxygen to breathe. It protects us from the sun's ultraviolet radiation, but it doesn't block the sun too much to prevent its energy from reaching Earth.

By absorbing some of the sun's energy, the atmosphere maintains a balance that keeps our planet's climate stable. If the atmosphere were any thicker, Earth would be too hot for life. If it were any thinner, our planet's surface would be vulnerable to more meteorite strikes. Furthermore, atmospheric pressure contributes to creating the just-right conditions for the earth to hold liquid water.

All these elements, along with countless other factors, are engineered necessities for life and are further evidence of the purposeful design found throughout God's creation.

Moon

While evolutionary scientists claim that Earth's moon was shaped by random processes, its features indicate that it was formed intentionally. In fact, life wouldn't be able to exist on Earth without it. Certain changes to the moon's location or size would disastrously affect our planet.

For example, the gravitational pull between our planet and the moon is a large cause of ocean tides. It stretches the waters on the sides nearest and farthest from the moon, creating high tides in those places and low tides everywhere else. The moon is also the right size and distance away to keep the tides balanced. If the moon were much bigger or closer to Earth, the

waters would cover larger portions of the continents and erode many low-lying landforms.

> "Do you not fear Me?" says the LORD. "Will you not tremble at My presence, who have placed the sand as the bound of the sea, by a perpetual decree, that it cannot pass beyond it? And though its waves toss to and fro, yet they cannot prevail; though they roar, yet they cannot pass over it." (Jeremiah 5:22)

The moon's gravity affects more than our seas—it keeps Earth's tilt stable. Without it, Earth would be prone to fluctuate on its axis. Its current angle of 23.5° creates seasons throughout the year as each hemisphere is positioned closer to or farther from the sun. If the moon didn't keep our planet's tilt stabilized, Earth could rotate either straight up and down (eliminating seasons) or on its side, like Uranus, placing one hemisphere in constant daylight for half the year and the other in night. The extreme temperatures caused by these changes would be detrimental for Earth's climate.

The moon's pull also keeps our planet from spinning too fast. If Earth spun much quicker, devastating weather would become more frequent, and increased waters would be drawn to the equator, potentially submerging entire countries.

Even the moon's rotation is perfectly synched with ours. It takes 27.3 days for the moon to rotate once—exactly the time it takes to orbit Earth. As a result, we consistently see the same side of the moon, even though it is spinning. This synchronous movement is known as tidal locking. All the large moons (and many of the smaller ones) in our solar system are configured this way.

Unlike other planets' moons, our moon provides the earth with one of the rarest sights of all—total solar eclipses. Though the sun's diameter is about 400 times larger than the moon's, it's also 400 times farther away from Earth. These exact proportions make them look the same size in our sky, allowing the moon to fully cover the sun during a total eclipse.

Sun

The sun is the powerhouse of our solar system. Its energy allows us to live on Earth. Yet, the sun itself is too hot for any object to draw close, much less sustain life. A toasty 27 million degrees Fahrenheit at its core, our sun is primarily made of hydrogen atoms that collide to form helium via nuclear fusion. This process releases light and heat into space and prevents the sun from collapsing on itself.

The sun is about 100 times wider than Earth, and this giant size provides the necessary pressure to maintain the sun's energy levels since so much is released into space. It also has enough gravity to securely hold the solar system's eight planets, along with several dwarf planets and thousands of asteroids and comets, in stable orbits. It's not too large, either—the sun is too small to explode into a supernova, so it's impossible

for the sun to ever become a neutron star (a star that collapses from using up all its fuel) or a black hole.

Even the sun's location in our galaxy is perfect for the planets in its orbit. It's found in a small spiral arm about midway between the galaxy's center and edge, while most stars and dangerous supernovas are nearer the middle. This perfect positioning keeps our sun at a safe distance from other stars, protecting us from nearby collisions and supernovas as well as high amounts of radiation.

Stars

In the night sky, we see only a small percentage of the stars that exist across our universe. They're found in a variety of sizes and colors, and are scattered throughout the far reaches of our galaxy and beyond. Of course, the sun is the nearest star to our planet, but even so, it's millions of miles away. We can't fathom just how many stars fill the skies or how far they extend.

The next closest star to Earth is Proxima Centauri. It's over 25 trillion miles away from our sun. These neighboring stars are only two of the 100 billion-plus stars in the Milky Way alone. Like the sun, many of these stars have solar systems of their own, all located at vast distances away from our planet. Even with the most advanced telescopes, we've barely skimmed the stars in our galaxy, much less those in the billions of other galaxies all over space.

> He counts the number of the stars; He calls them all by name. (Psalm 147:4)

Clouds

A cloud-covered eclipse is certainly disappointing, but those atmospheric "cotton balls" are important. Clouds cover about two-thirds of our planet's sky. They're created through the water cycle, a process in which water rises as a vapor into the air because of the sun's heat. Then, it turns back into droplets or ice crystals in the cool atmosphere to form clouds. When the clouds get too heavy, they release precipitation, nourishing dry land, maintaining moisture, and cooling the planet's surface before the cycle starts again.

Clouds also help regulate temperature, blocking the sun's rays to keep us cool and trapping heat to make us warm. There are several different types of clouds, each with its own formation. Some, like cumulus clouds, are white and fluffy. Others, like cumulonimbus clouds, can produce huge storms. These

kinds of clouds create lightning from electrical charges. Thunder is the sound produced by lighting when it strikes. Thankfully, all the cloud types have a distinct look that indicates what kind of weather we can expect. They're like natural forecasters.

Planets

For a space object to be classified as a planet, it must have gravity strong enough to hold it in a spherical shape, orbit a star, and have no other objects of similar size nearby unless

those objects are affected by its gravity. Some planets are made of rock, others are comprised mostly of gases, and still others are covered in ice. They can have dozens of moons in their orbits (and possibly more), and some planets even orbit around two stars at once.

No planet is exactly like another, which is another testimony to our God's creativity. There are eight planets in our solar system (sorry, Pluto), billions in our galaxy, and countless others across the universe. But not all can experience eclipses.

Several moving parts must be perfectly in place for an eclipse to occur. Each eclipse involves a star, a planet that orbits the star, and a moon that orbits the planet, but those are only the basic requirements. There are many factors, including distance, size, and alignment, that allow some planets to experience eclipses while others don't. Our Earth was designed to be an ideal observation spot—especially since it's the only planet with humans able to observe!

Asteroids, Comets, and Meteoroids

Asteroids are frequently known as "minor planets," but they're not true planets. In reality, these objects are relatively small, irregularly shaped rocks. They tumble through space in their orbit around the sun. Most of them are found in the asteroid belt between Mars and Jupiter, and their sizes range from several feet to hundreds of miles.

In contrast, comets are chunks of ice and dust with a solid core. Comets also orbit the sun, but the heat vaporizes them slightly when they come close, often forming bright tails. Usually, the nucleus of a comet is only a few miles wide, although its tail may be millions of miles long.

Finally, meteoroids are small rocky or metallic objects that orbit the sun. They're called meteors when they burn up in Earth's atmosphere, creating tails of light—also known as "shooting stars." Thanks to our planet's protective atmospheric covering, this debris usually breaks up into smaller particles known as meteorites by the time it reaches the ground, so there's little damage if it does get through our atmosphere.

Space

We can't begin to imagine the immensity of outer space. The farthest planet from the sun, Neptune, is nearly 2.8 billion miles away. That's like circling the Earth almost 112,000 times! Our solar system is huge, yet it represents only a small fraction of the universe's vast size—and it's only one of about 4,000 known solar systems in the Milky Way galaxy.

How big is our galaxy? For context, large distances in space are often measured in light-years, the calculated distance light travels in one year. One light-year is equivalent to approximately six trillion miles. The Milky Way is estimated to measure not one or two but 100,000 light-years wide.

Comet NEOWISE

One of the most mysterious objects that can be found in space is a black hole, a dense area of gravity that's so powerful nothing can escape—not even light. The Milky Way is believed to have anywhere from 10 million to one billion black holes. Thankfully, our solar system is a safe distance away from the nearest one, Gaia BH1, at 1,600 light-years from Earth.

The next-closest galaxy to our own is the Canis Major Dwarf Galaxy, located about 25,000 light-years from Earth. Beyond that, even more galaxies can be found, each thousands upon thousands of light-years away and spinning with its own stars, planets, and black holes.

Scientists don't know the exact number of galaxies in our universe. Several neighboring galaxies are categorized into groups and then into packs called clusters. These constitute enormous superclusters that are among the biggest known structures in space.

When you look up at the night sky and see a multitude of twinkling lights, consider the incredible complexity of space and just how little we know about everything that lies deep within our cosmos.

> When I consider Your heavens, thc work of Your fingers, the moon and the stars, which You have ordained, what is man that You are mindful of him, and the son of man that You visit him? (Psalm 8:3-4)

Crab Nebula

PART FOUR: ETERNAL IMPACT

The features of our vast universe demonstrate immeasurable complexity and order, including eclipses. The most popular explanation for the universe's origin tends to be the Big Bang model, but does the evidence support this? More importantly, what if these stunning displays point instead to a divine Creator who is much greater than the universe itself?

The Big Bang

In recent years, scientists have cited the Big Bang as the catalyst for our universe's existence. This theory claims that all matter resulted from a single point of dense energy (or "singularity") that rapidly began to expand, forming particles, atoms, and eventually the entire cosmos. Many researchers believe this model is credible because the universe is still expanding today—or so it appears.

In truth, the "proof" for this growth remains inconsistent. Even if the universe is expanding, who's to say this couldn't be intentional, established at the beginning to prevent the universe's gravity from collapsing on itself? And where did the necessary initial energy come from?

The issues don't stop there, either. There are several other problems with this secular explanation, including:

> **Antimatter**—The Big Bang should have resulted in equal amounts of matter and antimatter, but there is apparently very little antimatter in the universe.

Temperatures—The universe is claimed to be billions of years old, but if the Big Bang were true, there should be no large hot and cold areas left from the "explosion." But large-scale cold and hot spots exist.

Mature galaxies—According to conventional science, it takes billions of years for light from far-off galaxies to reach Earth. If we're seeing these galaxies as they were in the distant past, they should appear chaotic and disorganized since that's what they would've looked like after the Big Bang. Instead, even the farthest galaxies appear incredibly mature and complex.

Besides the lack of evidence, the theory itself is unreasonable. It's like spilling a bucket of paint and expecting the result to be Vincent van Gogh's *The Starry Night*. Of course, that would never happen. It takes an artist to create a masterpiece, just like it takes a Designer to make something as complex as our universe. Order doesn't come from chaos.

The Truth Is…

God created the universe. He provided all the components needed for eclipses—the sun, moon, and Earth. But beyond even these, our Creator structured the galaxy in such a way that our solar system can continue to exist. The moving parts outside our immediate family of celestial bodies are invaluable in holding the universe and all its elements together. If another star moved too close to our sun, we might see a cosmic collision. The precisely aligned orbits of planets in our solar system and others prevent them from spinning into one another. Even meteors are limited by Earth's design to mostly burn up in the atmosphere before reaching the surface.

Some people believe that the Big Bang could have served as a mechanism through which God established creation. Scientific contradictions aside, there are many issues with this line of thinking. First, the Big Bang demands a timeline of billions of years, but this is not supported by Scripture. There's nothing in Genesis (or elsewhere, for that matter) implying that the days of creation defined by each morning and evening were anything but seven literal days. Based on biblical history and even what we see in creation, our Earth is still young—in fact, it's only thousands of years old!

Second, the Big Bang model asserts that stars formed before the earth, but that's not what the Genesis record tells us. The Bible says that God made the earth on the first day of the creation week, while the sun, moon, and stars were established on the fourth.

Finally, the Bible confirms its own account of creation, even centuries after Genesis was written. Scripture tells us that in the last days, people will reject the biblical truth of our origins.

> For this they willfully forget: that by the word of God the heavens were of old, and the earth standing out of water and in the water, by which the world that then existed perished, being flooded with water. (2 Peter 3:5-6)

Not only is the scientific support lacking for the Big Bang, but the Bible clearly testifies that our universe was designed in six days by a sovereign God—not by a sudden moment of chance.

The Fall

Sadly, our universe is broken. As incredible as it currently is, it's not what it was originally meant to be. All creation began as "very good," complete and without blemish (Genesis 1:31). It wasn't until humans chose to disobey God that the world was cursed. Death became a reality. The once-perfect cosmos was

Did You Know?

Some people believe that the "darkness over all the earth" (Luke 23:44) at Jesus' crucifixion was the result of a solar eclipse. However, it was actually a supernatural event. Solar eclipses can only happen on a new moon, but Jesus was betrayed during the Passover week, which is celebrated around a full moon.

corrupted so that "the whole creation groans and labors," waiting to be "delivered from the bondage of corruption" (Romans 8:21-22).

People were affected as well. Every person sins and "fall[s] short of the glory of God" (Romans 3:23) as a result of the Fall. Now, all people are separated by sin from their good and holy Creator and Father.

Ever since then, people have searched for meaning in all the wrong places—including the cosmos. They've looked to the stars for guidance, worshiped celestial bodies, and even credited the universe itself for their existence. Rather than admiring the skies as the glorious work of their Creator, they turned to nature as an idol, even though it pales in comparison to its Maker.

Try as we might, we simply can't reach heaven on our own by turning to false gods or trying to be good enough for the true God. We'll always fail to meet God's perfect standard, and left to ourselves we'll continue to forsake our Creator. Without Him, there's no hope for eternity. It would take a miracle to change our fate.

The Greatest Love

Yet, God didn't abandon us. He gave us hope. He wants us to be reconciled with Him and have the abundant life that is found in Him alone. But His holiness demands justice. Someone had to take the punishment for our mistakes—and it would require a blameless sacrifice. Because He desired to have a relationship with us, God paid the ultimate price and offered Himself.

God sent His Son, Jesus Christ, to the earth. He lived the perfect life that we couldn't, proclaiming the good news of salvation. But His message wasn't always well received. Jesus was ultimately rejected and crucified by the people He created and loved. In His death, He took all our sins—past, present, and future—upon Himself. He endured the punishment that we de-

served, all out of His great love for us.

But death couldn't hold Him. Three days later, He rose from the grave! In His resurrection, Jesus conquered sin and death once and for all. When we put our trust in Him as Savior, God doesn't see our brokenness—He sees the righteousness of Christ. Now, all who confess their sins and place their faith in Him can have eternal life and a personal relationship with the God who created them.

Romans 8:38-39 tells us "that neither death nor life, nor angels nor principalities nor powers, nor things present nor things to come, nor height nor depth, nor any other created thing, shall be able to separate us from the love of God which is in Christ Jesus our Lord."

God's immeasurable love for us reaches beyond the solar system, the galaxy, and even the universe itself. Our world remains fallen, but God promises that He will one day return to make everything new. In the meantime, we can rejoice in the peace that comes from knowing our Savior, Jesus Christ. All creation may fail, but His faithfulness never will.

> "And lo, I am with you always, even to the end of the age." (Matthew 28:20)

Eclipse Connection

Like everything in this world, eclipses are temporary. They appear for a short time, only to vanish as quickly as they came. Even the astronomical components that allow them to happen will ultimately fade. The reality is that eclipses are beautiful works of creation, but they won't change your life. Only our Creator, Jesus Christ, has the power to change a person's heart and give eternal life.

We're blessed to see these marvelous sights alongside countless other wonders in the skies. All creation proclaims the glory of God, from the smallest particles to the brightest stars that are billions of light-years away from us. The next time you witness an eclipse, praise the Lord for His incredible artistry in every detail of our universe. And the Maker of our vast cosmos created you fearfully and wonderfully, too.

> Marvelous are Your works, and that my soul knows very well. (Psalm 139:14)

OUT OF THIS WORLD TRIVIA

How many inches does the moon move away from the earth each year? *Approximately 1.5 inches.*

Do eclipses make any sounds in space? *Nope. Space doesn't have air, which is needed to transmit sound waves.*

How many Earths could fit inside the sun? *About one million.*

What can't an astronaut do in space? *Burp.*

Who built the world's first reflecting telescope? *Isaac Newton.*

What are the two types of planets in our solar system? *Terrestrial and gas giants.*

What color are the hottest stars in space? *Blue.*

What did God create on Day 4 of the creation week? *Planets, stars, moons, and other celestial bodies.*

Which Greek astronomer predicted the first eclipse? *Thales.*

Which astronomer was inspired by an eclipse to study science? *Johannes Kepler.*

IMAGE CREDITS

Rehman Abubakr via Wikimedia: 10, 14, 19
Bigstock: 9
Callan Carpenter via Wikimedia: 4
ESO: 34
Geralt via Wikimedia: 39
Lsmpascal via Wikimedia: 26
NASA: 8, 11, 23-24, 28
Public domain: 30b
Kalpa Semasinghe via Wikimedia: 32
David R. Tribble via Wikimedia: 30t
Wikimedia: 13
Susan Windsor: 20-22

GLOSSARY

Baily's beads Small packets of sunlight that appear around the moon just before totality in a total solar eclipse.

Blood moon A phenomenon that describes the moon's dark reddish-orange appearance during a total lunar eclipse.

Corona The sun's outer atmosphere that can often be seen during a total solar eclipse.

Diamond ring effect An effect in which one final burst of sunlight is visible just before totality in a solar eclipse.

Light-year The calculated distance that light travels in one year, equivalent to approximately six trillion miles.

Lunar eclipse Astronomical event in which the earth moves between the sun and the moon, casting its shadow on the moon's surface.

Nuclear fusion The collision of two or more atomic nuclei to make a heavier one; for example, hydrogen that creates helium to generate the sun's energy.

Penumbra The outer region of an object's shadow (i.e., the shadow of the earth or moon).

Prominences Columns of gas that leap from the sun's surface, often a pinkish color.

Shadow bands Waves of light and dark that ripple across the ground during the first contact of a solar eclipse.

Solar eclipse Astronomical event in which the moon travels between the sun and Earth, throwing its shadow on parts of the earth.

Tidal locking The synchronous movement of a moon's rotation with its orbit around a planet.

Umbra The central area of an object's shadow (i.e., the shadow of the earth or moon).

RESOURCES FOR MORE INFORMATION

General Information

Eclipses, posted on solarsystem.nasa.gov
Moon in Motion, posted on moon.nasa.gov
Lunar Eclipses and Solar Eclipses, posted on spaceplace.nasa.gov
Solar System Planets, Order and Formation: A Guide, posted on space.com
Eclipse, posted on scienceclarified.com
Solar Eclipses and Lunar Eclipses, posted on eclipsewise.com.
What are the 5 Stages of a Total Solar Eclipse? posted on labmate-online.com

Observation Tips and Safety

Types of Solar and Lunar Eclipses, posted on timeanddate.com
Solar Eclipse and the Created Sun podcast, posted on ICR.org
How to View a Solar Eclipse Safely, posted on eclipse.aas.org
How to Photograph a Solar Eclipse, posted on bhphotovideo.com
Go See The Eclipse: A Glimpse of God's Glory by Chap Percival

Creation Science

Evidence from Nature, posted on ICR.org
That's a Fact: Big Bang? posted on ICR.org
Our Young Solar System, posted on ICR.org
Guide to the Universe by Institute for Creation Research
Creation Q&A: Answers to 32 Big Questions about the Bible and Evolution by Institute for Creation Research
Johannes Kepler: Exploring the Mysteries of God's Universe by Michael Stamp and Christy Hardy
Space: God's Majestic Handiwork
Earth: Our Created Home
Guide to Creation Basics
These books are available at ICR.org/store

RESOURCES FROM ICR.org/store

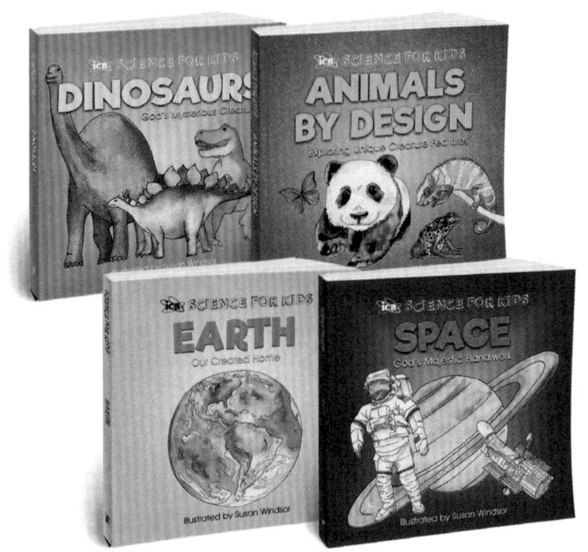

ABOUT THE INSTITUTE FOR CREATION RESEARCH

At the Institute for Creation Research, we want you to know God's Word can be trusted with everything it speaks about—from how and why we were made, to how the universe was formed, to how we can know Jesus Christ and receive all He has planned for us.

That's why ICR scientists have spent more than 50 years researching scientific evidence that refutes evolutionary philosophy and confirms the Bible's account of a recent and special creation. We regularly receive testimonies from around the world about how ICR's cutting-edge work has impacted thousands of people with Christ's creation truth.

HOW CAN ICR HELP YOU?

You'll find faith-building science articles in *Acts & Facts*, our bimonthly science news magazine, and spiritual insight and encouragement from *Days of Praise*, our quarterly devotional booklet. Sign up for FREE at **ICR.org/subscriptions**.

Our radio programs, podcasts, online videos, and wide range of social media offerings will keep you up-to-date on the latest creation news and announcements. Get connected at **ICR.org**.

We offer creation science books, DVDs, and other resources for every age and stage at **ICR.org/store**.

Learn how you can attend or host a biblical creation event at **ICR.org/events**.

Discover how science confirms the Bible at our Dallas museum, the ICR Discovery Center. Plan your visit at **ICRdiscoverycenter.org**.

P. O. Box 59029
Dallas, TX 75229
800.337.0375
ICR.org